Echo Auto Instructions Manual
Echo Auto User Guide
By Emery H. Maxwell

Text Copyright © 2019 Emery H. Maxwell

No part of this book may be reproduced
in any form or by any means without the
prior written permission of the author.

Disclaimer:
The views expressed within this book are those of the author alone. *Echo Auto* and *ALEXA* are trademarks of *Amazon*. All other copyrights and trademarks are properties of their respective owners. The information contained within this book is based on the opinions, observations, and experiences of the author and is provided "AS-IS". No warranties of any kind are made. Neither the author nor publisher are engaged in rendering professional services of any kind. Neither the author nor publisher will assume responsibility or liability for any loss or damage related directly or indirectly to the information contained within this book.

The author has attempted to be as accurate as possible with the information contained within this book. Neither the author nor publisher will assume responsibility or liability for any errors, omissions, inconsistencies, or inaccuracies.

Table of Contents

Introduction..........
How Echo Auto Works..........
Which Vehicles Echo Auto Connects to..........
    Vent Mount Compatibility..........
Getting Started..........
    Specifications..........
    External Display (Basic Hardware)..........
    Smartphones that are not Compatible with Echo Auto..........
    Light Bar Colors and their Meanings..........
How to Set up Echo Auto..........
    How to Download the Alexa App..........
    How to Set up Echo Auto with Bluetooth..........
    How to Set up Echo Auto with an Auxiliary Cable..........
How to Teach ALEXA Your Voice..........
How to Get Directions on Echo Auto..........
    Android..........
    iOS..........
How to Make Calls with Your Voice..........
    How to Use Your Voice to Send and Read Text Messages..........
How to Enable Skills..........
How to Use Echo with IFTTT..........
    How to Create an Applet..........
ALEXA Command and Request List..........
Troubleshooting..........
    How to Perform a Factory Reset of Your Echo Auto..........

Introduction

Welcome to the *Echo Auto Instructions Manual*. This user guide is intended to help you understand, set up, and manage *Amazon Echo* for your vehicle.

Amazon Echo Auto is essentially an *Alexa*-enabled device for vehicles. This device can be useful for the completion of tasks while you drive.

Driving accounts for a notable percentage of the average person's life. Many drivers are putting roughly 12,000 to 15,000 miles a year on their vehicles.

It's not surprising that people are interested in optimizing their driving experience.

With *Echo Auto*, you can simply ask *Alexa* to continue an *Audible* audiobook where you left off, make calls, play music, check your calendar, or locate nearby businesses. All of those things can be done in your vehicle.

The intention of this user guide is to provide an overview of what *Amazon Echo Auto* is capable of and how to get the most out of it.

This *Echo Auto Instructions Manual* will cover:

- **How *Echo Auto* works**
- **Light bar colors and their meanings**
- **A list of smartphones that are not compatible with *Echo Auto***
- **How to set up *Echo Auto***
- **How to get directions on *Echo Auto***
- **How to make calls with your voice**
- **How to use your voice to send and read text messages**
- **How to enable *Skills***
- ***Alexa* command and request list**
- **Troubleshooting**
- **and plenty more**

How *Echo Auto* Works

Echo Auto works by connecting to the *Alexa* app on your phone and uses your current smartphone data plan.

Simply plug the *Echo Auto* device into your vehicle's USB port and mount it to an air vent. For convenience, a vent mount is also available for the device.

The setup process is done through the *Alexa* app.

Echo Auto plays through your vehicle's speakers via auxiliary input or your smartphone's Bluetooth connection.

The device has 8 microphones that can pick up your voice commands, even when there is music, road noise, and air conditioning running in the background.

While it is designed for the road, it is also designed for privacy. You have the option of disconnecting the microphones whenever you like simply by pressing the microphone **Off** button.

Since it connects to the *Alexa* app on your phone, it can utilize *Google Maps* to help you navigate through areas on the road that you are unfamiliar with.

When the device is set up correctly, *Alexa* can complete many different tasks.

To activate *Alexa*, say, "*Alexa*." Then wait for the chime and ask a question.

Note: Since *Echo Auto* connects to the *Alexa* app on your phone and uses your current smartphone data plan, carrier charges may apply.

Which Vehicles *Echo Auto* Connects to

Y ou might be wondering whether *Echo Auto* is even capable of connecting to your vehicle, especially if you have not yet purchased the device yet.

The answer is it connects to the vast majority of cars that support Bluetooth to play music or that have an auxiliary input.

Here is a list of cars that are NOT compatible with *Echo Auto* via Bluetooth connection:

- Acura MDX (2016)
- Acura RDX (2017)
- Acura TLX (2016)
- Chevrolet Equinox ((2016, 2017)
- Chevrolet Malibu (2015, 2016, 2017)
- Chevrolet Silverado (2014, 2015, 2017)
- Chevrolet Volt (2015)
- Dodge Caravan (2018)
- Dodge Ram (2018)
- GMC Sierra (2015, 2016, 2017)
- Honda Accord (2013, 2015)
- Honda Accord Touring (2013)
- Honda Civic (2013)
- Honda CRV (2014)
- Honda Odyssey (2016)
- Honda Pilot (2013)
- Infinity Q60 (2018)
- Infinity QX60 (2019)
- Jeep Wrangler (2017)
- Mazda CX9 (2013)
- Mazda 3 (2010)
- Toyota Camry (2015)
- Toyota Corolla (2017)

Note: Although these vehicles are not compatible with *Echo Auto* via Bluetooth connection, they are still compatible via the included auxiliary cable, as long as your vehicle has an auxiliary input.

If you are still having difficulty connecting, there is something else wrong.

<u>Vent Mount Compatibility</u>

To verify that your vehicle's air vents are compatible with the included vent mount, check the diagram here.

If you are reading this publication in paperback form, the compatibility diagram can be found online.

To find the diagram online:

1.) Go to the *Amazon* website

2.) Search for *Echo Auto* in the search field

3.) Find the *Echo Auto* product and visit its sales page

4.) On the product's sales page, scroll down until you see the heading, **We want you to know**, and then click the **Check** link.

Getting Started

This section will cover the specifications of the device, the details of the external display, a list of smartphones that are not compatible with *Echo Auto*, and ring light bar colors and their meanings.

Specifications

Items included in the box: *Echo Auto* device, In-Car Power Adapter, Micro-USB cable (1m), 3.5mm auxiliary cable (1m), and Quick Start Guide.

Size: 3.3" x 1.9" x 0.5" (85 mm x 47 mm x 13.28) *Actual size may vary by manufacturing process*

Weight: 1.6 oz (45 grams) *Actual weight may vary by manufacturing process*

Processor: Mediatek MT7697, Intel Dual DSP with Inference Engine

Audio: Eight microphones. Supports auxiliary audio output.

Bluetooth: Hands-free Profile support for calling, Advanced Distribution Profile support for audio streaming, Audio/Video Remote Control Profile for voice control of media sessions, Serial Port Profile for Bluetooth connectivity to *Android* phones, and iPod Accessory Protocol for Bluetooth connectivity to iPhones.

Connectivity: *Echo Auto* connects to most cars that support Bluetooth to play music or that have an auxiliary input. Certain vehicles do not work well with *Echo Auto* via Bluetooth. If you are unable to connect devices to your car to play music via auxiliary cable or Bluetooth, you might be able to connect *Echo Auto* with an additional accessory, such as an FM transmitter or cassette tape adapter.

Smartphone plan usage and compatibility: *Echo Auto* uses your current smartphone plan along with the *Alexa* app for connectivity and other features. Carrier charges might apply. For information regarding any fees and limitations that apply to your plan, consult your carrier. *Echo Auto* supports *Android* 6.0 and *iOS* 12 or greater. However, not all smartphones are compatible with *Echo Auto*. For a list of smartphones that are not compatible with *Echo Auto*, see the *Smartphones that are not Compatible with Echo Auto* chapter.

External Display (Basic Hardware)

Light bar: The light bar is displayed across the front of the device.

Action button: This button is located on top of the device, on the right. It has a small dot at the center.

Microphone off: This button is located on top of the device, on the left. It has a circle with a line going through it.

Array of 8 microphones: The eight small gaps on top of device are where the microphones are.

3.5 mm audio output: This is located on the right side of the device.

Micro-USB power: This is located on the right side of the device, next to the audio output.

Smartphones that are not Compatible with *Echo Auto*

- Alcatel One Touch
- Alcatel Pixi 4
- BLU Tank Extreme & Extreme Pro 4.0
- BLU R2
- Google Nexus 4
- Google Nexus 6
- Honor 4C
- HTC Desire 610
- HTC M8
- HTC M9
- Huawei Honor 4C
- Huawei Mate SE
- iPhone
- iPhone 3G
- iPhone 3GS
- iPhone 4
- iPhone 4S
- iPhone 5
- iPhone 5C
- Kyocera DuraForce PRO (E6820)
- LeEco LePro 3
- Le Pro 3
- LG Tribute HD
- Samsung Galaxy J2
- Samsung Galaxy J3
- ZTE Blade A510

Light Bar Colors and their Meanings

Blue right to left: The device is powering up.

Blue left/right to center: Your request is being processed.

Blue side to side: The device is waiting for a Bluetooth connection.

Red: The microphones were turned off with the **Mic on/off** button on the device. To turn the microphones back on, simply press the **Mic on/off** button again on the device.

Purple: Setup failed.

Orange sweeping: The device is prepared for setup.

Orange pulses: Factory reset is in progress.

How to Set up *Echo Auto*

T he setup process consists of multiple steps that will be covered in this section.

How to Download the *Alexa* App

If you haven't done so already, download the *Alexa* app.

The *Alexa* app is compatible with:

• Fire OS 5.3.3 or higher

• Android 5.1 or higher

• iOS 11.0 or higher

If you have an *Alexa*-enabled *Fire* tablet, the *Alexa* app should already be on there, since it downloads onto it automatically.

Otherwise, you can download it from one of various places.

The *Alexa* app can be downloaded from:

• the *Amazon Appstore*

• the *Apple App Store*

• *Google Play*

After you reach the app store on your mobile device, simply search for "*Alexa app.*"

If you are using a desktop computer, the *Alexa* app can be downloaded at:

alexa.amazon.com

Note: Feature availability varies on the desktop app.

You will need to have the latest version of the *Alexa* app installed on your smartphone.

You can also check for updates by visiting the app store and searching for "*Alexa app.*" If you already have the app installed, your mobile device should recognize it, and if there is an update available, select **Update**.

How to Set up *Echo Auto* with Bluetooth

Before setting up *Echo Auto* with Bluetooth, verify that you have a compatible car stereo. You will need to have a car stereo that supports Bluetooth (4.0 or higher) and Bluetooth music playback.

1.) Turn on your car and set the stereo's input to **Bluetooth**.

2.) Enable Bluetooth on your smartphone.

3.) Open the *Alexa* app.

4.) Select **Devices**. Then select the icon that has a circle with a plus sign on it.

5.) Select **Add Device**.

6.) Select **Amazon Echo**.

7.) Select **Echo Auto**.

8.) Follow the on-screen instructions to finish setting up your device.

How to Set up *Echo Auto* with an Auxiliary Cable

1.) Plug in the device with included USB cable and power adapter. The car's built-in USB port can also be utilized.

2.) Plug in the included AUX cable to your car's auxiliary port and to *Echo Auto*.

3.) Turn on your vehicle and set the stereo's input to **AUX**.

4.) Enable Bluetooth on your smartphone.

5.) Open the *Alexa* app.

6.) Select **Devices**. Then select the icon that has a circle with a plus sign on it.

7.) Select **Add Device**.

8.) Select **Amazon Echo**.

9.) Select **Echo Auto**.

10.) Follow the on-screen instructions to finish setting up your device.

How to Teach *ALEXA* Your Voice

Teaching *ALEXA* to recognize your voice can help create a more personalized experience across certain supported features.

This can be achieved by creating a voice profile.

Create a Voice Profile

1.) On your *smart phone*, go to the menu and select *Settings*.

2.) Go to the *Accounts* section.

3.) Select **Your Voice**.

4.) Select **Begin**.

5.) Using the drop-down menu, select the device you want to interact with to teach *ALEXA* your voice.

6.) Select **Next**.

7.) When prompted, say the phrase out loud. Then select **Next** to go to the next phrase. You can also try the phrase again by selecting **Try Again**.

8.) Select **Complete**.

You should now see a confirmation page on the screen.

To confirm, ask *ALEXA*, "Who am I?" If the process went well, *ALEXA* will mention your name.

How to Get Directions on *Echo Auto*

One of the benefits of *Echo Auto* is that it has the ability to give you directions, which can be particularly useful if you get lost on the road.

Whether you are looking for directions to a particular place or you'd like turn-by-turn navigation, *Echo Auto* offers both of those options.

The process of setting up *Echo Auto* directions may differ somewhat depending on whether you are using an *Android or iOS*.

Android

Alexa uses the default navigation app on your smartphone and your data plan to offer directions to your destination(s).

Turn-by-Turn Navigation

To get turn-by-turn navigation, say, "Alexa, get directions to [destination]."

Home or Place of Business Navigation

If you'd like to ask *Alexa* to navigate to your home or place of business, go to **Your Locations** in the *Alexa* app settings and add the address.

Turn off Navigation

If you'd like to turn off navigation, say, "Alexa, cancel navigation."

iOS

1.) Go to the **settings** section on your mobile phone.

2.) Scroll down and select the **Alexa** app.

3.) From there, enable:

• Banner alerts

• Notifications

• Location permissions

4.) Say, "Alexa, get directions to [destination]."

5.) To open the navigation app on your smartphone, follow *Alexa's* instructions.

Home or Place of Business Navigation

If you'd like to ask *Alexa* to navigate to your home or place of business, go to **Your Locations** in the *Alexa* app settings and add the address.

Turn off Navigation

If you'd like to turn off navigation, say, "Alexa, cancel navigation."

Note: *Alexa* uses the navigation app on your smartphone and uses your data plan to provide directions.

How to Make Calls with Your Voice

To make a call using the *Echo* device, simply ask *ALEXA* to call the person you want to reach, mentioning the contact's name.

<u>To make a call to the *ALEXA* app or another *Echo* device, say:</u>

"ALEXA, call [person's name] Echo."

<u>To make a call to a landline or mobile number that is saved to your list of contacts, say:</u>

"ALEXA, call [person's name] on his/her home phone."

"ALEXA, call [person's name] mobile."

"ALEXA, call [person's name] at work."

"ALEXA, call [person's name] office."

<u>To verbally dial a mobile or landline without saying the person's name:</u> (Available on *Echo* devices only)

"ALEXA, call [say each digit, including the area code]."

<u>To control the volume, say:</u>

"ALEXA, turn the volume up / down."

You can also mute the line manually by using the **Microphone off** button on the device.

<u>To hang up, say:</u>

"Hang up."

"End call."

If you are making the call from the *ALEXA* app, you can also select the on-screen **End** tab to disconnect.

At this time, *ALEXA* does not support calls to certain types of numbers, such as:

• Emergency services

• Premium-rate or toll numbers

• International numbers outside of North America

• Dial-by-letter numbers

• Abbreviated dial codes

How to Use Your Voice to Send and Read Text Messages

You also have the option to send and read text messages by using your voice with *Alexa*.

Note: Text messaging is not supported on *iOS*.

To send and read text messages, say:

- "Alexa, send a text message to [number / name of contact]."
- "Alexa, read my text messages."

How to Enable Skills

Skills are voice-controlled capabilities that improve the *ALEXA* device's functionality. To illustrate, if you'd like *ALEXA* to tell you about specific upcoming events in your city, you would need to enable a specific skill for that.

Skills are what will allow you have *Alexa* tell you the odometer reading, tire pressure, etc.

Oftentimes, if you know the specific name of the skill you'd like to use, you can simply say, "*ALEXA*, enable [skill name]."

But sometimes certain skills need to be enabled through the *Amazon* website or the *ALEXA* app, while others might need to be activated by following the prompts from *ALEXA*.

Enable *Skills*

1.) Open the *ALEXA* app.

2.) Go to the menu and select **Skills**.

You can also go to the *Amazon* website and go into the *skills* section.

3.) Use the *search bar* to find a specific skill or browse through the skills by category.

4.) After finding the skill you'd like to use, select it to go to its detail page. The detail page should include at least one example of what to say to play or open the skill.

5.) On the skill's detail page, select **Enable Skill**.

Now you should be able to tell *ALEXA* to open the skill.

If you need help with the skill, say, "ALEXA, [skill name] help."

Manage *Skills*

1.) Open the *ALEXA* app.

2.) Go to the menu and select **Skills**.

3.) Select **Your Skills**.

4.) Select a *skill* to go to its detail page.

You should now see a list of available options.

Note: To find skills that are specific to *Echo Auto*:

A link is provided here.

If you are reading this publication in paperback form:

1.) Go to the *Amazon* store and set the search parameters to **Alexa Skills**.

2.) Leave the search field blank and select the search icon (magnifying glass icon). This will bring you to the **Alexa Skills** section of the store.

3.) Scroll down under the **Alexa Skills** section on the left part of the screen and select **Connected Car.**

Depending on what kind of car you drive and what kind of skill you choose to enable, *Alexa* may be able to:

- Start your vehicle

- Tell you how much gasoline is in your vehicle

- Lock or unlock doors

- Activate the horns and lights remotely

- Tell you what the tire pressure, fuel level, or oil life is

Scroll through the list of skills and read the description(s) to verify that you have found one that is compatible with your vehicle.

There are many different skills that are for very specific makes and models only, so read the description(s) on the web pages carefully.

The setup process will vary depending on what type of vehicle you drive, but many of the setup instructions can be found right on the description page of the skill.

How to Use *Echo* with *IFTTT*

IFTTT stands for *If This Then That*. It's an online service that uses rules (applets) to connect a variety of apps and devices together.

Generally, it helps users do more with their apps and devices.

The *ALEXA* device supports *IFTTT*, and it can trigger the *IFTTT* "rules" you have activated.

To illustrate:

If you ask *ALEXA* to find your phone, *IFTTT* can trigger your phone to ring. Or if you'd like have your *android* phone muted at bedtime, that can be done also.

New applets can be created, or you can choose from applets that already exist from other *IFTTT* users.

1.) If you haven't done so already, go to the [*IFTTT*](#) website and sign up.

2.) On the *IFTTT* website, find the *ALEXA* app by typing it into the search bar.

3.) Select the *Connect* tab when the *Amazon ALEXA* page comes up.

4.) Sign in to your *Amazon* account.

After signing in, your *Amazon* account should now be linked to your *IFTTT* account.

You have the option to remove the link between *ALEXA* and *IFTTT* at any time by visiting *Manage Login with Amazon*.

How to Create an Applet

1.) Click *invent your home* at the top-right corner of the screen on your *IFTTT* page.

2.) Select *New Applet*. Then click *+ This*. *If This* is the trigger part of the process.

3.) On the *Choose a service* page, type *ALEXA* into the search bar and select it.

4.) You should now be on the *Choose trigger* page. If you'd like to customize the wording, scroll down to the box that reads, *Say a specific phrase*. For example, to turn off the lights, you might want to say, "Power off," instead of "Turn off the lights," so that's what you would type in.

5.) After choosing a phrase to trigger the action, select *Create trigger*.

6.) Click on *+ That*. *If That* is the action part of the process. Then choose an action. For example, if you are using *WEMO* light bulbs and you'd like to power them on and off through *ALEXA*, search for *WEMO* in the search bar. In this case, you would then go to *WEMO lighting* and select *Connect*.

Although different apps have different connection methods, most of them are rather similar.

7.) After it's connected, go back and finish selecting an action. Following the previous example, if you have chosen *WEMO* light bulbs, you can now choose what you'd like to happen (Dim the light, Dim a group of lights, etc.) Then select the *Create action* tab.

8.) Select *Finish*.

To power the lights off, you would have to repeat the above steps. This time, on the *Create trigger* page, you can type in an "off" trigger, such as, "Power off."

There are plenty more things you can do, but the process generally remains the same. Scroll through the *IFTTT* lists and find something that interests you. When you are ready to set something up, follow the above steps outlined in this chapter.

ALEXA Command and Request List

Not everything in this list applies specifically to *Echo Auto*, but *Echo* in general. It is also not a definitive list, as there are tens of thousands of commands and requests.

Basics

"ALEXA, turn up the volume."
"ALEXA, turn down the volume."
"ALEXA, let's chat."
"ALEXA, stop."
"ALEXA, go to sleep."
"ALEXA, help."

Music

"ALEXA, next song."
"ALEXA, skip song."
"ALEXA, previous song."
"ALEXA, pause in [room name]."
"ALEXA, resume in [room name]."
"ALEXA, play the next track in [room name]."
"ALEXA, louder in [room name]."
"ALEXA, quieter in [room name]."
"ALEXA, set the volume to [volume number or percentage] in [room name]."
"ALEXA, mute [room name]."
"ALEXA, turn it up in [room name]."
"ALEXA, what's playing in [room name]?"
"ALEXA, play music by [artist]."
"ALEXA, what's this song?"
"ALEXA, buy [album name] by [artist's name]."
"ALEXA, play the top songs this week."
"ALEXA, play my [playlist name] playlist."
"ALEXA, shuffle my new music."

"ALEXA, shop for new music by [artist's name]."

"ALEXA, play the [station name] on [music service name]."

"ALEXA, add this song."

"ALEXA, who sings the song [song title]?"

"ALEXA, who is in the band [band's name]?"

"ALEXA, sample songs by [artist]."

To-do lists

"ALEXA, add [item] to my shopping list."

"ALEXA, create a to-do list."

"ALEXA, put [task] on my to-do list."

"ALEXA, I need to [task]."

Shopping on *Amazon*

"ALEXA, add [item] to my cart."

"ALEXA, buy [product]."

"ALEXA, order [item]."

"ALEXA, reorder [item]."

"ALEXA, where's my stuff?"

"ALEXA, track my order."

Smart Home

"ALEXA, discover my *smart home* devices."

"ALEXA, BLUETOOTH."

"ALEXA, connect to my phone."

"ALEXA, is the front / back door locked?"

"ALEXA, lock the front / back door."

"ALEXA, turn on the lights."

"ALEXA, turn on the TV."

"ALEXA, raise the temperature [number] degrees."

"ALEXA, set the temperature to [number]."

"ALEXA, what's the temperature in here?"

"ALEXA, what's the thermostat set to?"

"ALEXA, make the living room [color]."

"ALEXA, turn the desk lamp to [color]."

"ALEXA, turn on the hallway light."

"ALEXA, turn on *Movie Time*."

"ALEXA, dim the living room to [percentage]."

"ALEXA, set the fan to [percentage]."

Weather

"ALEXA, what's the weather in [name of city]."

"ALEXA, what's the temperature?"

"ALEXA, what will the weather be like in [name of city] tomorrow?"

"ALEXA, what's the extended forecast for [name of city]."

"ALEXA, is it going to rain today?"

"ALEXA, will it snow tomorrow?"

"ALEXA, will I need an umbrella today?"

Traffic and Local Information

"ALEXA, how is traffic?"

"ALEXA, what's my commute like?"

"ALEXA, what are the business hours of [venue name]?"

"ALEXA, what [venues] are nearby?"

"ALEXA, what time is the movie, [film name] playing?"

"ALEXA, find the address for [place]."

"ALEXA, is [venue] open?"

News

"ALEXA, what's in the news?"

"ALEXA, give me my flash briefing."

"ALEXA, open [publication name]."

"ALEXA, pause."

"ALEXA, next."

"ALEXA, previous."

Sports

"ALEXA, give me my sports update."

"ALEXA, what was the score of the [name of team] game?"

"ALEXA, did the [team's name] win?"

"ALEXA, when do the [team's name] play next?"

Alarm Clock

"ALEXA, set an alarm for [time]."

"ALEXA, when's my next alarm?"

"ALEXA, snooze."

"ALEXA, set a timer for [time length]."

"ALEXA, set a second timer for [time]."

"ALEXA, cancel my alarm for [time]."

"ALEXA, what time is it?"

"ALEXA, cancel all alarms."

"ALEXA, set a repeating alarm for [time] [days]."

Calendar

"ALEXA, what's the date?"

"ALEXA, add an event to my calendar."

"ALEXA, add a [time and event] to my calendar.

"ALEXA, what's on my calendar today?"

"ALEXA, what's my next appointment?"

Knowledge

"ALEXA, how tall is [name of mountain]?"

"ALEXA, how deep is [name of ocean]?"

"ALEXA, what's the capital of [place]?"

"ALEXA, what's the population of [place]?"

"ALEXA, who wrote [name of book]?"

"ALEXA, what's the definition of [word]?"

"ALEXA, how do you spell [word]?"

"ALEXA, what's [number] times [number]?"

"ALEXA, [number] factorial?"

AUDIOBOOKS

"ALEXA, play [book title] on *Audible*."

"ALEXA, pause."

"ALEXA, resume."

"ALEXA, next chapter."

"ALEXA, previous chapter."

"ALEXA, go to last chapter."

"ALEXA, go to [chapter number]."

Just for Fun

"ALEXA, tell me a joke."

"ALEXA, sing a song."

"ALEXA, tell me a story."

"ALEXA, play a game."

Troubleshooting

ALEXA App Doesn't Seem to Work

- Confirm your device meets the requirements
- Restart your phone
- If you are using a web browser, close the web browser, then reopen it
- Close the app, then reopen it
- Uninstall the app, then reinstall it

Problems with ALEXA Skills

- Disable the skill, then re-enable it

Music or Media Does Not Play with *Echo Auto*

First, verify that your car stereo is set to the correct input. Additionally, make sure that the volume on your car stereo and your phone is turned up high enough.

The solution to this issue will depend on your car's Bluetooth setup.

<u>Car with Bluetooth music and calling</u>

Solution: Set the input on your car stereo to Bluetooth. Verify that *Echo Auto* is connected to your phone over Bluetooth.

<u>Car with Bluetooth calling only</u>

Solution: Set the input on your car stereo to AUX. Verify that *Echo Auto* is connected to your car stereo via the included AUX cable.

<u>Car without Bluetooth</u>

Solution: Set the input on your car stereo to AUX. Verify that *Echo Auto* is connected to your car stereo via the included AUX cable.

Echo Auto Loses Bluetooth Connection

First, force the *Alexa* app to close, then restart your *Echo Auto*. If that doesn't resolve the issue, continue to the following troubleshooting steps.

Note: Check if *Echo Auto* is connected to your phone between each troubleshooting step.

1.) Unplug the micro-USB power cable from your *Echo Auto*. Then wait for 30 seconds and plug it back in.

2.) For *Android* phones: Turn on Airplane mode on your phone. Then wait for 45 seconds and turn it back off again.

For *iOS* phones: Turn off Bluetooth on your phone, then turn it back on.

3.) Restart your smartphone.

4.) Go to your *Echo Auto* settings in the *Alexa* app. From there, forget the device. Then go to your phone's Bluetooth settings to verify that your phone is connected to *Echo Auto*.

If your phone is in fact connected to *Echo Auto:*

a.) Unpair or forget the connection.

b.) Unplug the micro-USB power cable from your *Echo Auto*. Then wait for 45 seconds and plug it back in.

c.) Open the *Alexa* app and complete the setup process again.

5.) Complete a factory reset of *Echo Auto*, then complete the setup process in the *Alexa* app.

How to Perform a Factory Reset of Your *Echo Auto*

Note: A factory reset will erase your device settings.

1.) Press the **Mute** button, then hold down the **Action** button for 15 seconds until *Alexa* tells you that your device is resetting.

2.) Wait for the light bar to turn into a pulsing orange.

3.) Complete the setup process again.

www.ingramcontent.com/pod-product-compliance
Lightning Source LLC
Chambersburg PA
CBHW082026230526
45466CB00023B/3627